Eat the Storms

Eat the Storms

Damien B Donnelly

First published 2020 by The Hedgehog Poetry Press

Published in the UK by
The Hedgehog Poetry Press
Coppack House, 5
Churchill Avenue
Clevedon
BS21 6QW

www.hedgehogpress.co.uk

ISBN: 978-1-913499-26-6

9 8 7 6 5 4 3 2 1

A CIP Catalogue record for this book is available from the British Library.

Contents

Meditation Under the Yellow Sun

I wanted to draw
the sound of the moon
on a sun-drenched beach,
stripped down to white sand,
white wave, white skin
starved for affection.
I wanted to draw
the sound of that moon
as the chaos of the current
crashed upon crowds
clawing at each other
below a spot of sunlight
that burnt them quicker
than they could contemplate
contentment while I sketched
the white light of night
circling the circumference of day.

I wanted to put onto paper
that possibility of holding stillness
while all else moved;
of leaning into the moonlight melody
while the daylight drowned out thought,
of being able to hold silence in a song.

I wanted to draw the sound of the moon
as the yellow sun striped the sea from the sand.

Scarlet Rising

Eat the storms, Mother said,

boil these beds of bitter blackness
until the dream rips through the rain
and translucence turns to trust,

even a diamond must ache in the darkness
until compression can no longer conceal.

Eat the storms, Mother said,

slip out of shivering skin
until touch recalls
the sweet music of scarlet rising
caught below the lick of leaf
lost in the shadow of the shade,

even the petal must rise above the thorn
before it can dance in the light.

Eat the storms, Mother said,

but I didn't hear it, at first.

It takes time to swallow the truth
and teach the tongue to taste the rain.

The Purple Petal

Lather us in lazy,

let us lick honey
from purple petals,
let us lay down dreams
upon the velvet of a plump peach,
slip us into a dream of sleep
where language is lulled
into a lake of stilled thought
that tickles tongue upon first taste
with the truth of who we are,

where we shed red thorns
that have twisted flesh
and bequeath our blues
to the bed, at the bottom,
to form a base as we rise
in a garden of purple pride,

honey pouring
from our once-starved lips.

Black is Only Shadow

Winter has grey wings, feathers of soot
that come from concrete clouds
too dense to discern any light beyond.
Winter wears grey wings
but spring is an architect of possibility
by a canal of colour-
sweeping in after the frost to bathe us
in a fresh breath that blows across a chest
once in chains.

Round the red bricked bridge we ride,
each pedal pushing past the storms
that rained rivers through our winters.
Follow the river, she sings, *seasons are short
but the earth is a sphere turning towards the light.
dark doors open often into hopeful,
the river recalls its route regardless of the water,
blue can be a bright beacon, black is only shadow
before it finds a reason to ignite in light,
bark is dry but branches bear blossom.*

We can be the water or the bridge,
we can be the natural path or the paved plot.
The route is bright beyond the chains,
beyond the bend where the colour is waiting.

Ruby Red

We capture lies in jam jars,
rich ruby reds to dapple
sweetness over the bitter truth.

We fish through sieves
for reflections of who we were
before we drowned the earth dry.

We turn towards the clouds,
trying to draw conclusions
from those cotton constructions
we cannot catch hold of.

We walk across the world
with a faith

that can't always keep us afloat.

Catch Colour

The sun sets and rises
and in between we catch
the kisses that come upon the current
though the continent is not ours to conquer.

Tides come and go,
touch is temporary,
flesh is polished pink below the sky
but falls like sands in the glass
that hoards the hours,
like clouds that can never be caged.

The sun sets and we blaze
orange blossoms into passing nights,
the night's gale calls of connections
in the passing,
passion is precious until it too passes.

The sun rises and falls,
catch light; catch the fire
before it drowns on the water,
catch the colours to paint
the coming of the grey,
to keep afloat until the next kiss.

Catch colour,
catch kisses before the sun sets,
let worry waste upon the wave,
tomorrow's light will be blue enough.

Purple Clouds

In the many meadows of my mind

plants grow down
from purple clouds

of carved cotton

and seek substance
from the surface, not the ceiling.

In the many meadows of my mind

fences are painted
with faces familiar,

mouths catch kisses if you're quick enough

and embraces
sprout like bushes to cradle comfort.

In the many meadows of my mind

music spreads like ivy;
a chorus to cut through the chaos,

a crescendo of colour like a flower unfolding.

Grains of Sand Beneath Cerulean Skies

Faith is fragile,
courage is not always conclusive,
we don't command the waves
or comprehend the clouds.

I tell you this sand will be swept
into the sea by nightfall,
this baying breath of cyan
beneath the stretch of those cerulean skies.

This smooth, salt-licked land
was forged from fire before you were born,
when vultures had feathers
instead of hands and knives,
when volcanos were all there was to fear.

Faith is fragile,
we can't see what once was
or what will come to be,
we lie somewhere below the caelum
searching for security on a spot of shore
before the tides return and we,
in turn, become grains of sand
that some being will one day look upon
and try to see what is no longer there.

It is ours to be the basalt
or to be something better.

Shades of Blue

I hear you calling
in shades of blue
from the extremities
of a distance my arm's reach
can never cover.

I hear you calling
in shades of blue,
your concern comes in currents
across the continents,
in those cold corners
when I question creation
and my position within it.

I hear you calling
in shades of blue
only one born to know no origins
can truly discern
in these days
that demand ties
that have not been well tethered

The Greying Mist of Memory

I'd never heard colour call to me of green
though my eyes caress it in a certain light
and so many walls I've covered with that same colour
to curate a comfort from the cold.

Blue, I never found blue cold, on the contrary,
I see a sky come down to caress the seas I've crossed
in a coating of calm encouragement,
even in the snow light or moonlight,
blue light connects its contours
like icy jazz thrilling through a single saxophone
on a smoky soirée, in a time the greying mist
of memory hasn't quite drained.
Blue never, but white; now come the chills.

I had red walls once and thought them a tribute
to my, as yet unexposed, pride,
I now recall them as more melancholic;
a call in themselves, but, as a child,
I was scarlet conquering on Sunday afternoons
on the inside of the rain
as oldies played beneath the turntable's needle
long before I heard the blue in the singer's song.

Tattered Brown Trousers

Father ate all the flowers
in the back garden
because he couldn't swallow
the promise of happiness
blooming within the home
he couldn't find his root within.

Father left all the flowers
in the front garden,
too proud for others to see
him pulling from the soil
everything he needed help with
but had never been taught the words for.

Father liked to laugh, first,
when others lost,
so no one could hear his own loss
tearing at him, like weeds twisting
behind the restraints he wore
like his inside-out jumpers
and the tattered brown trousers
he thought no one could see through.

Father ate all the flowers
in the shadows
of the back garden
and choked on a laugh
that no one understood.

Rising Through the Rickety Reds

I laid on the floor and touched the marbled perfume
of the ocean as it washed over me, waves of flying feathers,
a fluid fire of salted foam. I kissed the poison of your lips, once,
and did not die as you came over me, next to me, inside of me.

Decay is not a breaking blue, not a pout of ruby red.

I have drowned more, before, on quiet corners, in safe seats,
in non-comforting crowds; dozens all searching
for their own spotlight in place of a single soul.

Spirit is often lost in too much light
and I wonder if the blind can see better?

The day can be a dark dance, fathers can decay in a garden
where hope can't win over weeds, mothers are sometimes made
beyond the cord that was cut from the blossom of another's labour,
bleeding can be a rite of passage like letting go, moving on.

Blood is not always the thicker bond and flowers can find a rhythm
in a rickety red room where the will is willing to wait and not be weighted.

We can't all be angels but we can rise upon the air we eat,
the touch we have tasted, the flesh we have crept from, swept upon,

found a fondness for,

even in rooms where naked blushes on the walls we have washed
with waves of a red raw hope that finds root in a simple light.

Golden Haze

The morning comes slowly,
eyes still dazzled by the delicate stars
now off trailing dust across the universe
as if plotting to tempt us further
than the stubborn stance of our spotlights
and I wonder how far you got as I sit here,
in this slowly waking morning light
casting shadows on the single form in this too big room
with no door large enough to climb through.

We considered setting sails
along cotton avenue once, long ago,
in a corner of this concrete jungle,
a single streetlamp casting courage onto our concerns
of cutting free like a jazz break from your base,
of burning our own trails of starlight
across the deafening daylight.

I am breath that still can bleed now, here now,
far from that corner we once we painted dreams on,
trying to force the foot to slow the speed
of this time burning while you; already taken to the dust,
now a speckled starlight cutting your own groove
into an orbit I can't observe
while tossing remembrances down from the night sky
that fall and flitter above the distraction
of this golden haze of mourning light,

still coming on slow.

Red Ink

I love and lose in circles, scratching at skin
tipped in ink, trying to find the truth
beneath the colours I've let others fill in,
returning to the paler flesh I kept from view.
We always need to hold on to something.

I am not comfortable over quiet dinners;
too much stilled air coursing over courses,
questioning the seconds ticking, in silence;

will you find me failure?

But I'll have taken flight first, a trick I learned
since that first separation I had no hand in.
Adoption is not an agreement signed by all parties.

I stir the stilled air with performances; shy boy
in the spotlight singing songs he can't quite
find the notes for, or the right to call his own.
Adoption and rejection are tied to separation.

I love and lie in circles that spiral back
on themselves, that cast further reflections,
not quite clear, on the boy now faced with a man
in the mirror, that flood more ink

into that fading flesh.

Variations grow stale, thought becomes tension,
creation becomes controlled, breath becomes
bearer, bleaker. My chest beats too quickly
to let in fresh air, fresh flesh, compressed, repressed.

I cannot lie in these circles, these spirals
that seem to linger. No longer. I am looking
to find a new shape; turning back,
recalling that first mark, to measure how far
from it I ran, to see what was left behind,
to lay it to rest and find the rest of me

beneath the red ink tipped into this flesh.

Grazing Greens

I set down
upon your shores,
those grazing greens
of my childhood memory
displaced as tears rain down
over the darkness
of your sleeping fields,

once seeping with humble hope,

once filled with a fine blood
even famine could not blight,

now flooded with a feeling
of regret or relief,

too dark to tell,
too changed to recognize,

not knowing if you are crying
because I found my way home
or that I'd once made a home on another shore.

Green Garden

Behold the moist moss
of early morning in green garden,
towering tree thriving through winter,

the peace that dawns with the dust,
the blue sky afloat on still water,

absorbing, reflecting, meditating,

the simple route the river runs,
the rustle of the red rose tipped with thorns,

the flowering moonlight over stony soil,
the secrets spring whispers to summer's stock.

The Irises of our Eyes

Crazed caught on canvas, caught in colour,
thought tempered in sweeping strokes-
we can be carried away in seas of grass,
greens awash in the garden, catch the canvas
before its fold finds favour in other fields
the mind has yet to fathom, we can be crazy.

Quick comes the crow upon the harvest,
bleak beacons, art is not always to be understood
nor the artist always allowed the freedom to express;
we want cream walls and canvases to comfort the couch,
expression doesn't always please the pattern.

Crazed comes to life on canvas,
see how he called to us; potato faced pickers
pealing in broken browns, aged in ochre,
acrylic is not a cover up and the canvas
not a vision of vanity, even the sunflowers
wilt before the irises of our eyes.
Fields of amber grain, far from home,
far from fame, trying to catch the elusive light
bearing down on the bails of honeyed hay
before the black wings hanging in the horizon.
Painting eyes, other's eyes for us to learn from,
to weep for the loss after the colour no longer connects.

Catch creation before it catches fire,
before it ricochets in a bed in Anvers-sur-Oise,
electricity only illuminated the intensity,
insanity is not always sedated by the shock.
Colour can't be captured in a brass bed
with brown leather straps.

Colour is conveyed on canvas, in connections,
in the bend the brush makes to blend,
in the waves the stars twist into that night sky,
in the lines of letters to brothers
who know us to be better
than the light sometimes allows.

He was a captive to the colour,
to the canvas, to the voices dark and distant,
cut it off and they still come a calling.

Capture colour before they caption you as crazy.

Yellow Light

ever knew
how far I could bend
before I would break,

till I snapped
before the sunrise,
before the yearning
the yellow light found me

ng,

looking for a lost breath
in the back of a dark chest
I had filled with every worry

that wasn't mine.

Even an elastic
knows its limit
before it lies limp,

before it can't recall
its own recovery,
before its tension
rips it from its reason.

At the Setting of the Yellow Light

I held your hand in a taxi, once,
while thinking of another
as you whispered into my ear
a sound I don't remember,
a scent now a breath away.

I can't hold everything anymore.

I recall the yellow light
yearning to hold its own innocence
stretching through the window,
burning hands still holding onto a truth
that had turned away from white.
I remember the highway that hurried us
out of the city as I wondered
if I'd packed enough hope for us both.

But I can't hold everything, anymore,
the elastic cannot be recoiled, the weight
was too wretched for just one heart.

I reach for that plant pot
with its purple petal planted, long ago,
in a garden I am returning to.

A garden where I will sit and watch
the dance of the dandelions
till the yellow sun has descended,
where I will empty all the jam jars
of their collected lies and draw
the sound of the moon, at last.

Colour is Waiting

And still we'll come to lick honey
from purple petals and still we'll come
to root out weeds of worthlessness
in gardens where others devour

all that is beautiful.

Time turns and we, in turn, follow its path,
suns set and the moon shows us its song.

Hold hands and then release.
Hold hope and then move on,
we only own the moment.

Mothers may still hand over their hearts
to other mothers waiting to be wanted,
fathers may rise to be fearless
or choke on the root of their own fear.

But remember; black is only shadow.

Still we come to that lake where language
lingers, where we sink beneath its depths
to slip from reflections we once wore
and have long since outgrown.

Come, catch the colour, catch the kisses,
catch the life racing by in taxis, on trains,
under starry nights waiting for us to paint them.
Behold how much there is to love,
to let go of, to learn from.

Eat the storms, she said. Eat the Storms.

"This is a richly-suggestive, highly-vivid collection of poems, depicting a wide canvas of emotional landscapes with painter-like precision. The intensely visual aspect of Donnelly's writing is encapsulated in 'The Irises of our Eyes', where the poetic speaker is caught up in a visionary experience, where nature, art and language become one.

'Crazed caught on canvas, caught in colour,
thought tempered in sweeping strokes-
we can be carried away in seas of grass'

It's no surprise that Donnelly is also a painter and designer and repeated references to colour amplify and intensify emotions in rhapsodic waves of word-pictures, provoking intense feelings of joy and grief, as seen in the poignant 'Tattered brown trousers', where the fragilities of a father are darkly-depicted.

'Eat the Storms' is a sensory reading experience, with an accessible, appealing and multi-layered voice. Each new reading reveals different shades of meaning and all the nuances of Donnelly's lyric voice. At once tender and lyrical, there are poignant and troubling moments throughout: the pain of experience - 'the simple route the river runs, the rustle of the red rose tipped with thorns'; the recurring motif of fragile relationships; the conflicting desires for belonging and freedom; the gut-wrenching theme of being deserted; the complexities of identity and the ever-shifting sense of self we experience - 'beneath the red ink tipped into this flesh'.

This is a striking, powerful collection, which achieves a balance between a personal, expansive and lyric style and the taut control needed to achieve fine poetry."

Matthew M. C. Smith
Poet and editor of Black Bough Poetry.

"Damien Donnelly is the archduke of alliteration and a poet in love with colour and sensuality. The poems in this first collection pop with blinding whites, rich reds and purples, and the yellow of Van Gogh sunflowers. They are jazz riffs on journeys, prisms chasing and catching the light. There is darkness too, in lines such as "Father ate all the flowers/in the back garden", and there are slower poems of self-assessment, including this finely achieved passage in "Red Ink":

"I am looking
to find a new shape; turning back, returning,
recalling that first mark, to measure how far
from it I ran, to see what was left behind..."

Above all, there is a sweeping positive energy, a welcoming of all that the world has to offer, and a certainty that "dark doors open often into hopeful".

Catherine Ann Cullen
Poet

"In a pamphlet saturated in colour, Damien Donnelly takes us on an immersive journey through a landscape of pigments. Written with great lyricism and emotional intensity, these poems contrast darker hues with lighter tones to create a sequence of poems that will linger in the memory."

Jessica Traynor
Poet

Acknowledgments

This would never have been possible without love and support of my mother and all the Donnelly clan along with the encouragement of my tribe of literary lumières who have inspired, cheered and pushed me out into the light and colour, including the Wordpress gals Jane Dougherty, Merril D. Smith, Sarah Connor, Liz Cowburn and Stefanie Neumann and the later homeland additions of Kevin Bateman, Eilín de Paor, Aishling Keogh, Mari Maxwell and, for their gracious time given to writing the blurbs; Catherine Ann Cullen, Jessica Traynor and Matthew M C Smith. Finally, a huge thank you to Mark for giving these poems a home at Hedgehog and for welcoming me into the hoglet family. Forever grateful to you all.